SPACE TRAVEL
from Then to Now

BY RACHEL GRACK

AMICUS | AMICUS INK

Sequence is published by Amicus and Amicus Ink
P.O. Box 1329, Mankato, MN 56002
www.amicuspublishing.us

Copyright © 2020 Amicus. International copyright reserved in all countries. No part of this book may be reproduced in any form without written permission from the publisher.

Library of Congress Cataloging-in-Publication Data
Names: Koestler-Grack, Rachel A., 1973- author.
Title: Space travel from then to now / by Rachel Grack.
Description: Mankato, MN : Amicus/Amicus Ink, [2020] | Series: Sequence Developments in Technology | Includes bibliographical references and index. | Audience: Grade 4-6.
Identifiers: LCCN 2018048883 (print) | LCCN 2018049520 (ebook) | ISBN 9781681517698 (pdf) | ISBN 9781681516875 (library binding) | ISBN 9781681524733 (pbk.)
Subjects: LCSH: Manned space flight--Juvenile literature. | Outer space--Exploration--Juvenile literature.
Classification: LCC TL793 (ebook) | LCC TL793 .K61174 2020 (print) | DDC 629.4/1--dc23
LC record available at https://lccn.loc.gov/2018048883

Editor: Wendy Dieker
Designer: Aubrey Harper
Photo Researcher: Holly Young

Photo Credits: KOHb/iStock cover; Official SpaceX Photos/Flickr cover; Eric Van Den Brulle/The Image Bank/Getty 4; Rolls Press/Popperfoto/Getty 6–7; NASA 9, 20–21; Herbert Haseneder/Flickr 12–13; NASA/WikiCommons 10–11, 14, 16–17, 18; Pixabay 22; NASA – JPL 24–25, 26–27, 28–29

Printed in the United States of America

HC 10 9 8 7 6 5 4 3 2 1
PB 10 9 8 7 6 5 4 3 2 1

TABLE OF CONTENTS

The First Trips to Space	5
To the Moon and Beyond	8
Stations and Shuttles	15
Robots in Space	23
Into the Future!	28

■ ■ ■ ■ ■

Glossary	30
Read More	31
Websites	31
Index	32

Sputnik's four antennas sent radio data to Earth. It opened space travel possibilities.

The Soviet Union launches Sputnik into space.

OCTOBER 4, 1957

...LOADING...LOADING...

The First Trips to Space

Space travel was once something just for storybooks. People looked up at the sky. What could be out there? Could vehicles fly beyond the sky? Could people travel to space? On October 4, 1957, a rocket lifted Sputnik into space. This Soviet **satellite** was the first spacecraft. Since then, space travel keeps blasting farther into the universe!

Sputnik's launch started what's known as the **Space Race**. The Soviet Union and the United States both wanted to explore space. Who would make it first? In April 1961, the first person flew into space. Soviet cosmonaut Yuri Gagarin **orbited** Earth. Then he landed safely.

> Yuri Gagarin spent 108 minutes flying around Earth in a ship called *Vostok 1*.

The Soviet Union launches Sputnik into space.

OCTOBER 4, 1957

APRIL 12, 1961

Soviet Yuri Gagarin becomes the first person in space.

To the Moon and Beyond

In the 1960s, the Space Race was in full force. Soviet and U.S. crafts were flying to space. **Pilotless** crafts were even flying around the Moon. But it wasn't until 1968 that astronauts circled the Moon. **NASA** sent a three-man crew on the Apollo 8 mission. They orbited the Moon 10 times. What a view of Earth!

The Soviet Union launches Sputnik into space.
OCTOBER 4, 1957

Soviet Yuri Gagarin becomes the first person in space.
APRIL 12, 1961

Apollo 8 sends the first people around the Moon.
DECEMBER 21, 1968

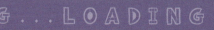

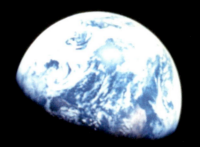

This famous photo is called "Earthrise." Men on the Apollo 8 mission took it while orbiting the Moon.

| The Soviet Union launches Sputnik into space. | Apollo 8 sends the first people around the Moon. |

OCTOBER 4, 1957 — APRIL 12, 1961 — DECEMBER 21, 1968 — JULY 20, 1969

| Soviet Yuri Gagarin becomes the first person in space. | U.S. astronauts land on the Moon for the first time. |

Could a person actually land on the Moon? Yes. The Apollo 11 mission launched on July 16, 1969. Four days later, U.S. astronauts Neil Armstrong and Buzz Aldrin stepped onto the Moon's surface. They planted an American flag. They gathered Moon rocks to study when they got back to Earth.

Neil Armstrong is in the reflection on Buzz Aldrin's mask during the first trip people took to the Moon.

The galaxy had much more to see. In 1972, NASA sent *Pioneer 10* to go exploring. This was the first **space probe** to fly beyond Mars. It reached Jupiter. And it kept going. In 1983, it left the **solar system**. In 2003, the signal became too weak to hear. Today, it could be billions of miles away.

An artist's computer drawing shows what *Pioneer 10* looks like as it travels through outer space.

The Soviet Union launches Sputnik into space.	Apollo 8 sends the first people around the Moon.	*Pioneer 10* heads for Jupiter.
OCTOBER 4, 1957	DECEMBER 21, 1968	MARCH 2, 1972
APRIL 12, 1961	JULY 20, 1969	
Soviet Yuri Gagarin becomes the first person in space.	U.S. astronauts land on the Moon for the first time.	

...OADING...LOADING...LOADING...

Astronaut Owen Garriot controls some tools in the Skylab.

The Soviet Union launches Sputnik into space.	Apollo 8 sends the first people around the Moon.	*Pioneer 10* heads for Jupiter.	
OCTOBER 4, 1957	APRIL 12, 1961	DECEMBER 21, 1968	JULY 20, 1969	MARCH 2, 1972	MAY 14, 1973
	Soviet Yuri Gagarin becomes the first person in space.		U.S. astronauts land on the Moon for the first time.		Skylab launches into orbit.

Stations and Shuttles

What if astronauts could live and study in space? In 1973, the Skylab space station was sent into space. Rocket ships carried three crews to the space station and back home again. The final crew lived there 84 days. At the time, that had been the longest stay in space. In 1979, Skylab fell back to Earth.

Early space ships were built for just one flight. NASA wanted to build a space shuttle that could fly again and again. They did it! The shuttle *Columbia* lifted off on April 12, 1981. It landed two days later. In the next 30 years, NASA space shuttles carried more than 350 astronauts on different space missions.

Columbia is lifted into space by a rocket. The rocket will fall away and *Columbia* will fly on its own in space.

The Soviet Union launches Sputnik into space.		Apollo 8 sends the first people around the Moon.			*Pioneer 10* heads for Jupiter.	
OCTOBER 4, 1957	APRIL 12, 1961	DECEMBER 21, 1968	JULY 20, 1969	MARCH 2, 1972		MAY 14, 1973
	Soviet Yuri Gagarin becomes the first person in space.		U.S. astronauts land on the Moon for the first time.		Skylab launches into orbit.	

APRIL 12, 1981

LOADING... LOADING...

Columbia lifts off on the first space shuttle mission.

| The Soviet Union launches Sputnik into space. | Apollo 8 sends the first people around the Moon. | *Pioneer 10* heads for Jupiter. |

OCTOBER 4, 1957 — APRIL 12, 1961 — DECEMBER 21, 1968 — JULY 20, 1969 — MARCH 2, 1972 — MAY 14, 1973

| Soviet Yuri Gagarin becomes the first person in space. | U.S. astronauts land on the Moon for the first time. | Skylab launches into orbit. |

These new shuttles carried more than just people. They took tools to space, too. A shuttle carried the Hubble Space **Telescope** in 1990. Scientists use this telescope to see the far reaches of space. The Hubble continues to discover galaxies we can't see from Earth.

> **Astronauts bring new parts to the Hubble on a repair mission.**

The Hubble Space Telescope is sent to space.

APRIL 12, 1981   APRIL 24, 1990

Columbia lifts off on the first space shuttle mission.

After the Skylab project ended, astronauts wanted to study in space again. In 1998, space programs began building the International Space Station (ISS). It was a joint project among several countries. It was launched into space that year. Groups of scientists have lived there since 2000. They stay for months at a time.

> Astronaut Tracy Caldwell Dyson looks at Earth from windows in the International Space Station.

The Soviet Union launches Sputnik into space.		Apollo 8 sends the first people around the Moon.		Pioneer 10 heads for Jupiter.	
OCTOBER 4, 1957	APRIL 12, 1961	DECEMBER 21, 1968	JULY 20, 1969	MARCH 2, 1972	MAY 14, 1973
	Soviet Yuri Gagarin becomes the first person in space.		U.S. astronauts land on the Moon for the first time.		Skylab launches into orbit.

| APRIL 12, 1981 | APRIL 24, 1990 | NOVEMBER 20, 1998 | ...LOADING... |

The Hubble Space Telescope is sent to space.

Columbia lifts off on the first space shuttle mission.

The first parts of the International Space Station are sent to space.

21

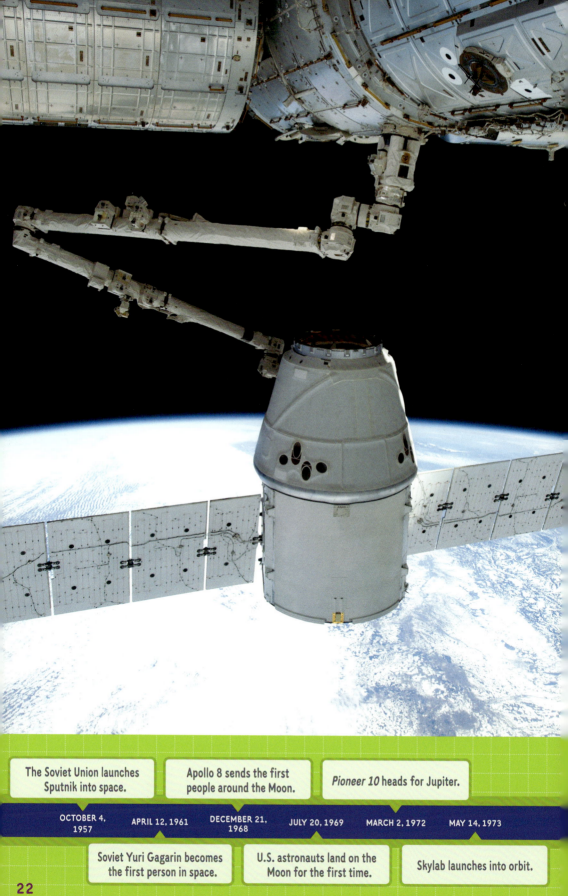

The Soviet Union launches Sputnik into space.	Apollo 8 sends the first people around the Moon.	*Pioneer 10* heads for Jupiter.	
OCTOBER 4, 1957	APRIL 12, 1961	DECEMBER 21, 1968	JULY 20, 1969	MARCH 2, 1972	MAY 14, 1973
	Soviet Yuri Gagarin becomes the first person in space.		U.S. astronauts land on the Moon for the first time.		Skylab launches into orbit.

Robots in Space

Spacecraft deliver parts and supplies to the ISS. But not all of them have people inside. In 2012, SpaceX sent its *Dragon* capsule to the ISS with some **cargo**. The *Dragon* uses radar to find its way to the ISS. Astronauts in the ISS watch it connect by itself. They take over with remote control if they need to.

The *Dragon* connects to the ISS to bring supplies.

APRIL 12, 1981 — *Columbia* lifts off on the first space shuttle mission.

APRIL 24, 1990 — The Hubble Space Telescope is sent to space.

NOVEMBER 20, 1998 — The first parts of the International Space Station are sent to space.

MAY 22, 2012 — SpaceX sends its pilotless *Dragon* spacecraft to the ISS.

Space probes have gone to space since the early days of space travel. They send data back to Earth. Today, a **rover** named *Curiosity* explores Mars. This car-sized robot landed on Mars on August 6, 2012. NASA scientists send it instructions. Then it travels around Mars by itself, gathering information to send back to Earth.

Curiosity rolls around the surface of Mars, gathering information to send to Earth.

The Soviet Union launches Sputnik into space.

Apollo 8 sends the first people around the Moon.

Pioneer 10 heads for Jupiter.

OCTOBER 4, 1957 — APRIL 12, 1961 — DECEMBER 21, 1968 — JULY 20, 1969 — MARCH 2, 1972 — MAY 14, 1973

Soviet Yuri Gagarin becomes the first person in space.

U.S. astronauts land on the Moon for the first time.

Skylab launches into orbit.

APRIL 12, 1981	APRIL 24, 1990	NOVEMBER 20, 1998	MAY 22, 2012	AUGUST 6, 2012
Columbia lifts off on the first space shuttle mission.	The Hubble Space Telescope is sent to space.	The first parts of the International Space Station are sent to space.	SpaceX sends its pilotless *Dragon* spacecraft to the ISS.	Exploration rover *Curiosity* lands on Mars.

| The Soviet Union launches Sputnik into space. | Apollo 8 sends the first people around the Moon. | *Pioneer 10* heads for Jupiter. |

| OCTOBER 4, 1957 | APRIL 12, 1961 | DECEMBER 21, 1968 | JULY 20, 1969 | MARCH 2, 1972 | MAY 14, 1973 |

| Soviet Yuri Gagarin becomes the first person in space. | U.S. astronauts land on the Moon for the first time. | Skylab launches into orbit. |

Rovers and probes send information about the surface of Mars. But what lies deep below? Is the inside of the planet like Earth? To find out, a robot probe was sent to drill into Mars. *InSight* was launched May 5, 2018. It landed November 26, 2018. Scientists are eager to learn even more about Mars.

A drawing shows *InSight* drilling into the surface of Mars.

APRIL 12, 1981 — *Columbia* lifts off on the first space shuttle mission.

APRIL 24, 1990 — The Hubble Space Telescope is sent to space.

NOVEMBER 20, 1998 — The first parts of the International Space Station are sent to space.

MAY 22, 2012 — SpaceX sends its pilotless *Dragon* spacecraft to the ISS.

AUGUST 6, 2012 — Exploration rover *Curiosity* lands on Mars.

NOVEMBER 26, 2018 — *InSight* probe lands on Mars.

Into the Future!

Since 1957, scientists have worked to get people farther into space. Could *people* actually travel as far as Mars? Scientists say yes. NASA and the European Space Administration are building a new spacecraft. They call it *Orion*. In 2020, a mission around the Moon will test the craft. Someday *Orion* may carry people to Mars!

An illustration shows what *Orion* might look like as it flies in space.

OCTOBER 4, 1957	The Soviet Union launches Sputnik into space.
APRIL 12, 1961	Soviet Yuri Gagarin becomes the first person in space.
DECEMBER 21, 1968	Apollo 8 sends the first people around the Moon.
JULY 20, 1969	U.S. astronauts land on the Moon for the first time.
MARCH 2, 1972	Pioneer 10 heads for Jupiter.
MAY 14, 1973	Skylab launches into orbit.

APRIL 12, 1981	APRIL 24, 1990	NOVEMBER 20, 1998	MAY 22, 2012	AUGUST 6, 2012	NOVEMBER 26, 2018
Columbia lifts off on the first space shuttle mission.	The Hubble Space Telescope is sent to space.	The first parts of the International Space Station are sent to space.	SpaceX sends its pilotless *Dragon* spacecraft to the ISS.	Exploration rover *Curiosity* lands on Mars.	*InSight* probe lands on Mars.

Glossary

cargo A load of supplies and tools carried by a spacecraft.

NASA Short for National Aeronautics Space Administration; the part of the U.S. government that is in charge of studying space.

orbit To fly in a circle around a planet, moon, or other space body.

pilotless A spacecraft or aircraft that does not have a pilot on board.

private To do with a nongovernment organization.

rover A wheeled vehicle used to explore another planet.

satellite A spacecraft that orbits, or circles, a planet or moon.

solar system The Sun and the planets, moons, and other bodies that orbit it.

space probe A pilotless spacecraft that explores space and sends data back to Earth.

Space Race The time when U.S. and Soviet scientists were working hard to be the first ones to explore different areas of space.

telescope An instrument that makes distant objects look larger or closer.

Read More

Rice, Earle Jr. *The Orion Spacecraft.* Hallandale, Fla.: Mitchell Lane Publishers, 2018.

Spray, Sally. *Awesome Engineering Spacecraft.* Mankato, Minn.: Capstone Press, 2018.

VanVoorst, Jenny Fretland. *Spacecraft.* Minneapolis, Minn.: Pogo, 2017.

Websites

European Space Agency for Kids
www.esa.int/kids/en/home

NASA Kids Club
www.nasa.gov/kidsclub/index.html

Our Universe for Kids
www.ouruniverseforkids.com

Every effort has been made to ensure that these websites are appropriate for children. However, because of the nature of the Internet, it is impossible to guarantee that these sites will remain active indefinitely or that their contents will not be altered.

Index

Aldrin, Buzz 11
Apollo missions 8, 11
Armstrong, Neil 11
Columbia 16
Curiosity 24
Dragon 23
Gagarin, Yuri 6
Hubble Space Telescope 19
InSight 27
International Space Station (ISS) 20, 23
Jupiter 12
Mars 12, 24, 27, 28
Moon 8, 11, 28
NASA 8, 12, 16, 24, 28
Orion 28
Pioneer 10 space probe 12
probes 12, 24, 27
rovers 24, 27
satellites 5
shuttles 16, 19
Skylab 15, 20
Space Race 6, 8
SpaceX 23
Sputnik 5, 6

About the Author

Rachel Grack has worked in children's nonfiction publishing since 1999. Rachel lives on a small desert ranch in Arizona. She enjoys spending time with her family and barnyard of animals. Thanks to our wireless world, her ranch stays tapped into developing technology.